Fun With Mathematics

Sumita Bose

Manda
Publishers
.com

Title: Fun with Mathematics

Author: Sumita Bose 2022

First Published by Manda Publishers in 2022

ISBN: 9789395174114
Price in INR: 275/- INR
Manda Publishers
www.mandapublishers.com
publish@mandapublishers.com
+91 9999190496

Cover Design: Tanya
Typography: Siddharth

Distributed by:
Amazon, Flipkart, Manda Publishers etc.

Printer:
Thomson press
New Delhi

Content

Friend not Foe

The subject 'Mathematics'

creates a lot of fear,

If the concepts are cleared

it can bring you some cheer.

The secret to score well

is to practice everyday,

Learn all the tables and

keep your fear at bay.

Make 'Maths' your friend

and not your foe,

Since you need it daily

to keep life on the go.

Introduction

Mathematics is a significant part of life. Right at birth our life gets surrounded by numbers. The doctor writes our weight, length, date and time of birth. Thus begins our journey with Mathematics and lasts throughout our life.

Sometimes we see that the children are not comfortable with Mathematics. The word 'Mathematics' instils a lot of nervousness. A nervous child grows into a nervous adult if the fear is not removed. The aim of this book is to remove the fear and make Mathematics an enjoyable subject.

Read the instructions attentively. Don't hesitate to take help from parents or grandparents if required. Do not feel upset if you don't get the desired result the first time. Try again. Have fun!

Chapter 1
Mathemagic

a) Dice Magic

Materials required:
Three dice, paper, pencil

Magic Show:
1) Ask your spectator to roll three dice.
2) Without lifting the dice, you can tell the sum of the numbers at the bottom of the three dice.

Trick:
Add the numbers on top of the dice and subtract the sum from 21.

Example:
If the dice shows 6, 3 and 2 the sum of the bottom numbers will be 10 (6 + 3 + 2 = 11, 21 − 11 = 10).

Note:
The above magic can also be shown with a single die or two dice. In case of a single die subtract the number

on the top from 7. In case of two dice subtract the sum of the numbers on top from 14.

b) Thought Reader

Materials required:
Paper, pencil, a book

Magic Show:
1) Write the magic number 2997 on a piece of paper, fold it and keep it inside a book. Do not show the magic number to your spectator.

2) Ask your spectator to write a three-digit number with three different digits.

3) Now you write a three-digit number below the spectator's three-digit number.

4) Next ask him/her to write another three-digit number below your number.

5) Again, you write a three-digit number.

6) Repeat steps 2 and 3.

7) You can write the sum of six numbers in just a few seconds.

8) Ask your spectator to open the folded paper.

9) Your spectator will be surprised to see the answer.

Trick:

a) The main trick lies in the three-digit numbers which you write. To get your three-digit number each time you subtract your spectator's three-digit number from 999 (subtract every digit from 9).

b) The answer will always be 2997 for six three-digit numbers. Don't show this magic to the same person multiple times.

Example:

```
     456 ← Spectator's number
     543 ← Your number (999 − 456 = 543)
     219 ← Spectator's number
     780 ← Your number (999 − 219 = 780)
     107 ← Spectator's number
   +892 ← Your number (999 − 107 = 892)
   ─────
    2997 Sum
```

Note:

The above magic can also be done with four or two three-digit numbers. For four three-digit numbers the magic sum will always be 1998 and for two three-digit numbers it will be 999.

c) Calendar

May 2022

SUN	MON	TUE	WED	THU	FRI	SAT
1	2	3	4	5	6	7
8	9	10	11	12	13	14
15	16	17	18	19	20	21
22	23	24	25	26	27	28
29	30	31				

Materials required:

Calendar, pen or a pencil, paper

Magic Show:

1) Select any month of the calendar.

2) Ask your spectator to choose any three successive (following in order) dates.

3) Now ask him/her to add the three dates and tell you the sum.

4) Just by knowing the sum you can tell the three dates.

Trick:

Divide the sum of three dates by 3. The answer will give you the middle date. Subtract 1 from the middle date to get the first date. Add 1 to the middle date to get the third date.

Example:

1) Suppose your spectator selects 29, 30, 31.

2) 29 + 30 + 31 = 90.

3) Middle date = 90 ÷ 3 = 30, first date = 30 − 1 = 29, third date = 30 + 1 = 31.

d) Consecutive Numbers

Materials required:

Paper, pen or a pencil, a book

Magic Show:

1) Write the magic number 3087 on a piece of paper, fold it and keep it inside a book. Do not show the magic number to your spectator.

2) Ask your spectator to write any four-digit number with successive (one after the other) digits in decreasing order.

3) Ask him/her to reverse the order of the digits and subtract it from the original number.

4) Now ask him/her to open the folded paper.

5) Your spectator will be amazed to see the answer.

Trick:

The answer is always 3087. Don't show this magic to the same person multiple times.

Example:

a) 1, 2, 3; 4, 5, 6 etc. are consecutive/successive digits.

b) Suppose your spectator chooses 8765.

c) Reversed number will be 5678.

d) 8765 − 5678 = 3087.

Note:

The above magic can also be done with three-digit numbers with successive digits in decreasing order in a similar manner. In this case the answer will always be 198.

e) Multiplication Race

Materials required:

Paper, pencil, calculator (optional)

Magic Show:

1) Ask your spectator to write a three-digit number and write it twice leaving a gap in between.

2) Next ask him/her to write another three-digit number below the first three-digit number.

3) Now you write a three-digit number below the second three-digit number.

4) Now ask him/her to find both the products and add. Challenge him/her to do it faster than you.

5) You will always be a winner in this race even without using a calculator.

Trick:

a) The main trick lies in the three-digit number which you write. Subtract your friend's second three-digit number from 999.

In the given example, 999 − 409 = 590.

b) Now subtract 1 from your friend's first three-digit number. In the above example, 783 − 1 = 782. This is the first part of the answer.

c) The end part of the answer is calculated by subtracting the first part from 999. In the above example, the last part of the answer is 999 − 782 = 217.

d) Thus, the final answer is 782217.

Example:

783← Spectator's number   783← Spectator's number

×409← Spectator's number   ×590← Your number

Chapter 2

Puzzles

a) Geometriku

In the given grid draw cube, cuboid, cone and cylinder in such a way that each row, each column and set of four squares touching each other uses every picture only once.

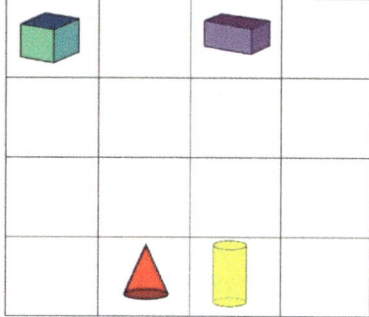

b) Shape number

In the given table each shape represents a number. They add up to the number on the right side of the box. Find the sum of a square, triangle and a circle.

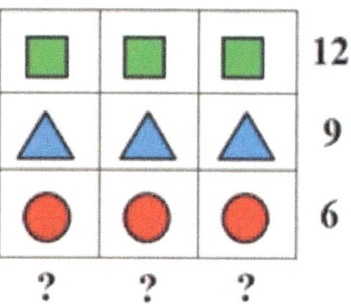

c) Clock Time

The following digital clocks show time in some pattern. Find out the pattern and tell the time of the fourth clock.

d) Three Numbers

Which three numbers give the same answer when they are multiplied as when they are added?

e) Weight Measurement

Which is heavier? One pound of feather or one pound of stones? (1 kilogram = 2.2 pound)

f) Drying Clothes

On a dry sunny day grandma Neena, hung out her grandson's clothes to dry on a clothesline. It took one hour for a shirt to dry completely. How much time would have been taken if she hung ten shirts?

g) Match Sticks

Rearrange three match sticks in such a way that you get five triangles.

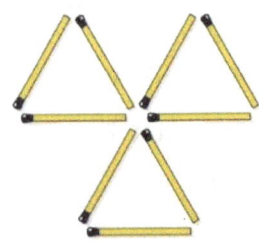

h) Buying Stamps

There are 12 one cent stamps in a dozen. Runa wants to buy 20 cent stamps. How many of them will she get in a dozen? (Dollar is the currency of USA, 1 dollar = 100 cents)

i) Shapes in Flags

Name a country for which the national flag contains the following shapes –

a) Red circle

b) Red semi-circle

c) Blue squares and rectangles

d) White stars (greater than 40)

e) Green right-angled triangle

f) Red trapezoid

g) Red parallelogram

j) Honeycomb

What is the shape of the honeycomb?
Why do the bees select this shape?

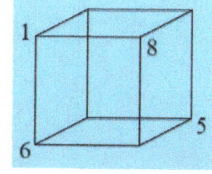

k) Cube

Write numbers 2, 3, 4 and 7 in the corners (vertices) of

the cube in such a way that each face adds up to 18.

l) Digital Clock

Each numeral on a digital clock is made up of seven, line segments. At what time of the day will the most segments have lit up on the clock?

m) Line Segment

Draw a line segment of length 5cm. Without using an eraser how can you make this line segment shorter.

n) Line Sum

Draw a straight line on the magic square in such a way that the sum of the numbers through which the line passes is 31.

4	3	8
9	5	1
2	7	6

Chapter 3

Optical Illusions

What is an optical illusion?

Optical illusion is a visual illusion. In an optical illusion our eyes see an image which does not exist (in reality). See the following images carefully and answer the given questions.

a) Müller Lyer illusion

This illusion was devised by Franz Carl Müller-Lyer, a German sociologist (an expert in the study of development, structure and functioning of human society) in 1889.

Which line is longer?

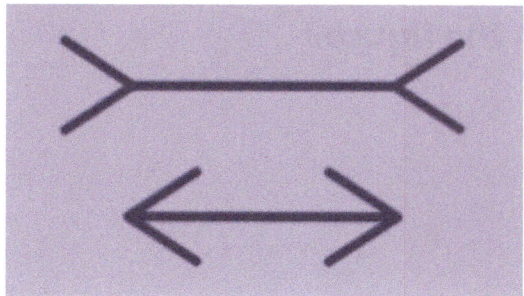

b) Zöllner illusion

This illusion was discovered by German astrophysicist (scientist who specializes in studying space, stars, planets and the universe) Johann Karl Friedrich Zöllner in 1860.

Are the thick lines parallel or intersecting?

c) Kanizsa illusion

This illusion was created by Italian psychologist (an expert in the study of human mind, emotions and behavior) Gaetano Kanizsa in 1955.

What shapes do you see in the figure?

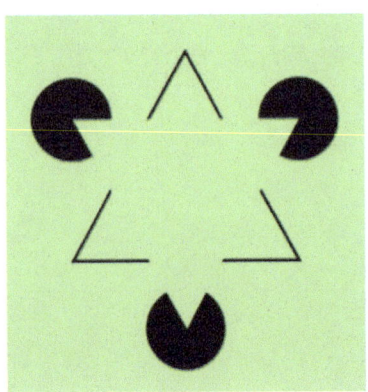

d) Hering illusion

This illusion was discovered by the German physiologist (an expert in the study of the normal functions of living organisms and their parts) Ewald Hering in 1861.

Are the two vertical red lines straight or bowed outward (at the centre)?

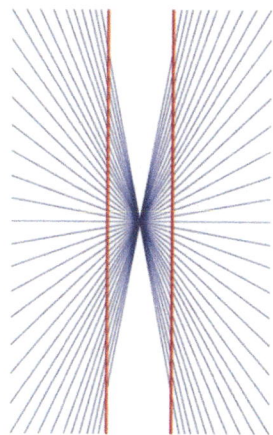

e) Circle illusion

How many big circles are there? Now bring your nose towards the red circle and count the number of big circles.

How many did you see?

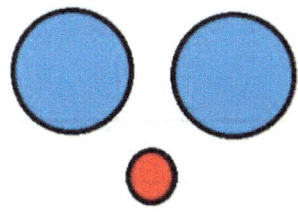

Chapter 4

Engrossing Patterns

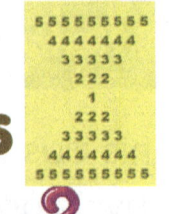

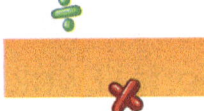

1) Observe the patterns and write the next line.

a)

$$4 \times 4 = 16$$
$$34 \times 34 = 1156$$
$$334 \times 334 = 111556$$
$$3334 \times 3334 = 11115556$$

b)

$$7 \times 9 = 63$$
$$77 \times 99 = 7623$$
$$777 \times 999 = 776223$$
$$7777 \times 9999 = 77762223$$

2) Observe the patterns and write the next two lines.

a)

$$9 \times 2 = 18$$
$$99 \times 2 = 198$$
$$999 \times 2 = 1998$$
$$9999 \times 2 = 19998$$

b)
$$11 \times 11 = 121$$
$$111 \times 111 = 12321$$
$$1111 \times 1111 = 1234321$$
$$11111 \times 11111 = 123454321$$

_____

_____

3) Observe the patterns and write the next three lines.

a) $\frac{1}{9} = 0.11111 \ldots$

$\frac{2}{9} = 0.22222 \ldots$

$\frac{3}{9} = 0.33333 \ldots$

$\frac{4}{9} = 0.44444 \ldots$

$\frac{5}{9} = 0.55555 \ldots$

_____

_____

_____

b) $\frac{1}{11} = 0.09090909...$

$\frac{2}{11} = 0.18181818...$

$\frac{3}{11} = 0.27272727...$

$\frac{4}{11} = 0.36363636...$

$\frac{5}{11} = 0.45454545...$

_____

_____

_____

4) Observe the patterns and write the next four lines.

a) $15873 \times 7 = 111111$

$15873 \times 14 = 222222$

$15873 \times 21 = 333333$

$15873 \times 28 = 444444$

_____

_____

_____

_____

b) $3 \times 37 = 111$ and $1 + 1 + 1 = 3$
$6 \times 37 = 222$ and $2 + 2 + 2 = 6$
$9 \times 37 = 333$ and $3 + 3 + 3 = 9$
$12 \times 37 = 444$ and $4 + 4 + 4 = 12$

_____

_____

_____

_____

Chapter 5
Activities

a) Birthday Magic Card

Learning objective:
To make a birthday magic card.

Materials required:
Card stock, ruler, pencil,
markers/sketch pens

Important Terms:

Diagonal of a square – A line segment which joins the opposite corners (vertices) of a square. It is not horizontal (–) or vertical (|).

Birthday magic square is a 4 × 4 {4 rows (→) and 4 columns (↓)} square containing a particular date, month and year in such a way that the rows, columns and diagonals add up to the same number. The formula for birthday magic square was created by the famous Indian Mathematician Srinivas Ramanujan between 1887 and 1919.

Method:

1) Fold the card stock into half.

2) Draw a 4 × 4 square on top of the card.

3) Use the following table to fill in the numbers.

d = date, **m** = month, **Y** = first two digits of the year,

y = last two digits of the year

d	m	Y	y
y+1	Y−1	m−3	d+3
m−2	d+2	y+2	Y−2
Y+1	y−1	d+1	m−1

4) Write happy birthday below the table on the cardstock and your birthday message inside the card.

Example:

Let us consider the date 14th November 2019.

Substituting the values d = 14, m = 11, Y = 20, y = 19 in the table given in step 3 we get:

14	11	20	19
20	19	8	17
9	16	21	18
21	18	15	10

The magic sum is 64 (sum of every row, every column and the two diagonals is 64).

Exercise:

Make a card for your friend, parent, teacher, uncle, aunt or grandparent using a magic square.

b) Measurement

Learning objective:
To estimate and find the actual length of various body parts.

Materials required:
Tape measure (measuring tape), paper, pencil or pen

Important terms:
Estimation means a rough calculation of the value i.e., an educated guess.
Circumference is the distance around the object.

Method:

Arm span

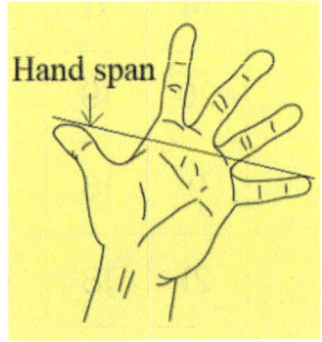

Hand span

1) Estimate the measurements of body parts and objects as mentioned in the table given below.

Body Part	Estimated measurement	Actual measurement
Your sibling or mom's hand span		
Your sibling or mom's arm span		
Your sibling or mom's forearm (wrist to elbow)		
Your sibling or mom's height (tell them to lie down)		
Length of your study table		
Breadth of your study table		
Circumference of any bottle		

2) Now use a measuring tape to find the actual measurements.

3) Record your observations.

4) Fill in the blanks using your observation table -

a) The handspan is close to the_ _ _ _(height/forearm).

b) The arm span is close to the _ _ _ _ (forearm/height).

c) The length of the table is _ _ _ _ than the breadth (greater/less).

Exercise:

Estimate the length and breadth of your bedroom then verify it by actual measurement.

c) Area & Perimeter

Learning objective:
To find the area and perimeter of the age (number) on the grid paper.

Materials required:
Grid paper, pencil, markers or crayons

Important terms:

Area is a measure of the space contained within an object. **Perimeter** is the distance around the object.

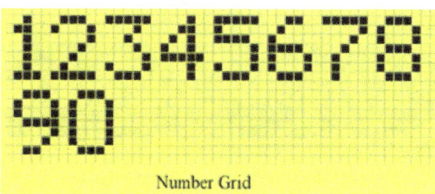

Number Grid

Method:

1) Write your age on the above grid.
2) Color the squares which are included inside the boundary of your age number.
3) Count the total number of squares inside the boundary of your age number. This is the area.
4) Count the number of line segments (side of each square) forming the boundary of the number. This is the perimeter.
5) Record your observations.

Observation:

Total number of squares of my age = Area = _ _ _ _ _ _
Square units.

Total line segments of the boundary of my age
= Perimeter = _ _ _ _ _ _ units.

Exercise:

Find the area and perimeter of your house number or phone number on grid paper.

d) Fraction, Decimal and Percentage

Learning objective:

To express colored squares of a given grid pattern in terms of fraction, decimal and percentage.

Materials required:

Grid paper, ruler, pencil, markers/sketch pens

Important term:

Percent means out of 100.

Method:

1) Count the total number of squares in the grid paper.
2) Count the number of Yellow, Brown, Blue, Red, Green and

White squares and note them in the given observation table.

3) Now write the fraction, decimal and percent of each color.

Observation:

Total number of squares = _ _ _ _

Serial Number	Color	Number of squares	Fraction	Decimal	Percent

Exercise:

Draw your own pattern on a 10 by 10 grid and record your observation table for the various colors.

e) Rangoli on Dot Paper

Learning objective:

To make a Rangoli pattern on a dot paper.

Materials required:

Dot paper, ruler, pencil, markers/sketch pens

Important term:

Rangoli is an art form originating in India (an Asian country) in which various patterns are created on the floor using rice powder, coloured powder, flowers etc. It is normally made during festivals.

Method:

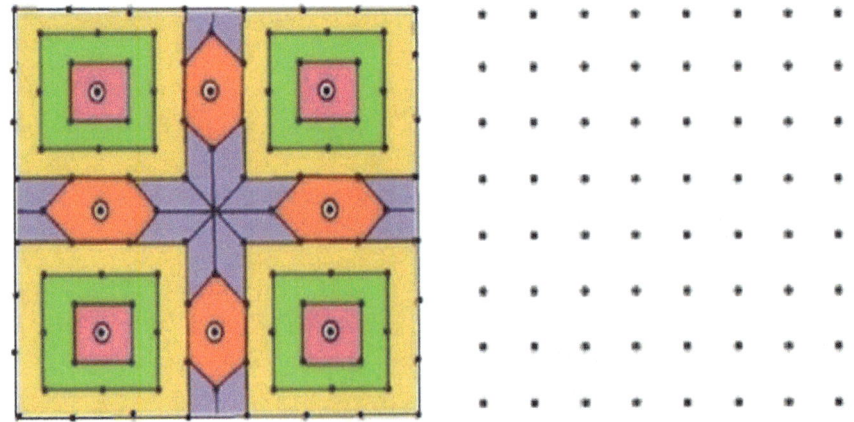

1) Draw the above rangoli pattern on the dot paper.

2) Start from any corner.

3) Join the dots using a ruler.

4) Colour the rangoli pattern.

Draw your own design on a dot paper (If you don't have a dot paper you can make one by putting dots on a paper at equal distances).

f) Day from a date

Learning objective:
To find the day from a date (any date from 1600 to 2199).

Materials required:
Card stock, ruler, pencil, markers

Important term:
In terms of calendar a **century** is a period of 100 years.

Method:
1) Use a cardstock to make the day code, month code and century code tables.

Day	Sun-day	Mon-day	Tues-day	Wednes-day	Thurs-day	Fri-day	Satur-day
Code	1	2	3	4	5	6	0

Month	Code	Month	Code
January	1	July	0
February	4	August	3
March	4	September	6
April	0	October	1
May	2	November	4
June	5	December	6

Century codes are all even numbers.

Century	Code
1600 – 1699	6
1700 – 1799	4
1800 – 1899	2
1900 – 1999	0
2000 – 2099	6
2100 – 2199	4

2) Find the sum by using the formula –

Sum = Date + Month code + Century code + Year + (Quotient of year ÷ 4)

3) Divide the sum by 7.

4) The remainder will be the code for the day (Look up in the day code table).

Note:

1) In the given formula by year we mean the last two digits of the year. Example, for a date in 2022 you must take the year to be 22.

2) Whenever you have any date of January or February of a leap year then code for the day is obtained by subtracting 1 from the remainder.

Example:
Find the day of 15 August, 1947 (Declaration of independence of India).

Solution:
Date = 15, Month code for August = 3, Century code for 1947 = 0, Year = 47 (Last two digits of 1947),
Quotient of year ÷ 4 = 47 ÷ 4 = 11
Using the formula we get–
Sum = Date + Month code + Century code + Year + (Quotient of year ÷ 4)
Sum = 15 + 3 + 0 + 47 + 11
Sum = 76
In 76 ÷ 7 Quotient = 10, Remainder = 6
6 is the code for the day.
Day code 6 corresponds to Friday.
Thus 15 August 1947, was a Friday.

Exercise:

Which day is December 25, 2175?

g) Mathematics A to Z

Learning objective:
To self-evaluate in order to find out the weak topics in Mathematics and take action to make them strong.

Materials required:
Paper, pencil or pen, ruler.

Important term:
Self-evaluation means self-assessment i.e., to look at your own progress and determine what areas have improved and what still needs improvement.

Method:
1) Draw four columns and give headings as letter, topic, emoji and action required.
2) In the first column write letters A to Z in each row respectively.
3) In the second column write Mathematics topics starting with the given letter in the first column. All the letters may not have any Math topic. A single letter may have more than one topic.
4) In the third column use the following emoji's as per your assessment.

Super hard	Hard	Moderate	Easy	Very easy

5) In the fourth column write the action required for improvement.

Example:

Letter	Topic	Emoji	Action required
A	Area	😐	Practice more problems
B	Bar graph	🙂	Help your classmates who find it hard
C	Carry over (regrouping in addition)	🙂	Try to do challenging problems
D	Division	☹️	Take help from teacher/parent

Exercise:

Make your own self evaluation table for Mathematics and Science.

Chapter 6
Amazing Facts

a) The Loop of 99

1) Write any three-digit number with different digits in ones and hundreds place.
2) Reverse the digits of the above three-digit number.
3) Subtract the reversed three-digit number from the three-digit number chosen in step1.
4) Reverse the obtained answer and subtract it from the previous answer.
5) If we continue the process of reversing and subtracting, we will always end up at 99.

Example:

a) Write any three-digit number with different digits in ones and hundreds place	871
b) Reverse the digits of the above three-digit number	178
c) Subtract the reversed three-digit number from the three-digit number chosen in the first step.	871−178 = 693
d) Reverse the obtained answer and subtract the smaller three-digit	693−396 = 297

number from the bigger three-digit number	
e) Repeat step 'd'	792−297 = 495
f) Continue the process	594−495 = 99

b) Strange Date

On 5th June 1978 at 12:34 there was a strange sequence of numbers − 12345678 (12:34, 5/6/78 date/month/year format). This can happen again only in 2078.

c) Alphabetical Order

The letters of the number 'forty' are arranged in the alphabetical order while the letters of the number 'one' are arranged in the reverse alphabetical order. There are no other numbers like these.

d) Number Strokes

Letter	Number of strokes
T	2
W	4
E	4
N	3
T	2
Y	3

N	3
I	1
N	3
E	4
Total	29

In English language, 'TWENTY-NINE' is the only number that is written with the same number of strokes as its numerical value.

e) Repeated Age

1) Multiply 259 by 39.

2) Now multiply the result of step one by your age.

3) Your age will repeat three times. (It works for every age less than 100). For ages less than 10 the result is separated by zeros.

Example:
a) Let the age be 12.

b) Multiply 259 by 39 $259 \times 39 = 10101$

c) Multiply 10101 by 12 $10101 \times 12 = 121212$

f) The 777 Number Trick

1) Write any number between 500 and 1000.

2) Add 777 to the selected number.

3) Remove one from the thousands place of the sum obtained in step '2'.

4) Add one to the remaining three-digit number.

5) Subtract the sum obtained in step '4' from the original chosen number.

6) You will always end up in 222.

Example:

a) Any number between 500 and 1000	938
b) Add 777	938 + 777 = 1715
c) Remove one from thousands place	̶1715
d) Add one to step 'c'	715 + 1 = 716
e) Subtract step 'd' from the original chosen number	938 − 716 = 222

g) Palindrome

A palindrome is a word, phrase, number or sequence of words that reads the same forwards as backwards. For example, mom, dad, radar, 12321 etc.

111,111,111 × 111,111,111

= 12,345,678,987,654,321

h) Anagram

An anagram is a word, or a phrase formed by rearranging the letters in another word or phrase. For example, the word 'cat' can be formed by rearranging the letters of the word 'act'.

Not only does 12 + 1 = 11 + 2 but the letters 'twelve plus one' re-arrange to give 'eleven plus two'.

i) Alphabets in Numbers

Letters A, B, C, D do not appear in the spellings of 1 to 99 in English language. Letter 'A' appears for the first time in the spelling of thousand, 'B' in billion, 'D' in hundred and 'C' does not appear.

j) Friendly Nine

If you multiply a number by nine and add all the digits of the product the sum will always be divisible by nine.

Example:

a) Multiply a number by nine	$87 \times 9 = 783$
b) Add all the digits of the product	$7 + 8 + 3 = 18$
c) Sum of the product is divisible by nine	$18 \div 9 = 2$

k) Odd Number

The spelling of every odd number in English language has an 'E' in it.

l) Mirror Image

73 is the 21st prime number. The mirror of 73 is 37 which is the 12th prime number.

m) Rubik's Cube

A 3 × 3 × 3 Rubik's cube has 43,252,003,274,489,856,000 possible combinations.

n) Sum and Product

First number	Second number	Sum	Product
9	9	9 + 9 = 18	9 × 9 = 81
3	24	3 + 24 = 27	3 × 24 = 72
2	47	2 + 47 = 49	2 × 47 = 94
2	497	2 + 497 = 499	2 × 497 = 994

o) Two-digit number ending in 9

Every two-digit number having 9 in the ones place is sum of the product of the two digits and the sum of the two digits.

Example:

$39 = (3 × 9) + (3 + 9)$

$89 = (8 × 9) + (8 + 9)$

Chapter 7
Tricks and Short Cuts

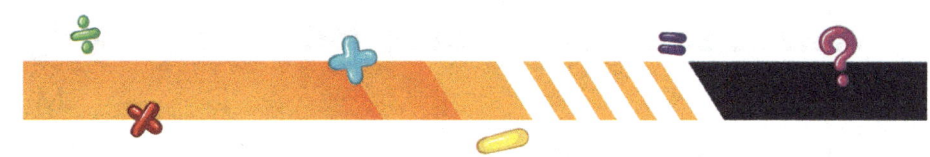

Symbols, tables, mnemonic (a learning technique in which a word, sentence or a poem is used for easy understanding and memorization of a rule/formula).

a) Place Value

To recall the correct order of the place value table just remember the sentence - "Mr. Harsh Thukral told Thiya and Thody he tasted octopus".

Mr.	Harsh Thukral	told Thiya &	Thody	he	Tast-ed	Octo-pus
Milli-on	Hund-red Thou-sands	Ten Thou-sands	Thou-sands	Hund-reds	Tens	Ones
M	H Th	T Th	Th	H	T	O

b) Roman Numerals

To recall the seven letters of Roman numeral, remember the sentence - "I view Xenia, Lily, Carnation, Daisy and Marigold".

I	view	Xenia	Lily	Carnation	Daisy	Marigold
I	V	X	L	C	D	M
1	5	10	50	100	500	1000

c) Comparison

<, > are Alligator mouths. The mouth opens towards the bigger number.

d) Expanded Notation

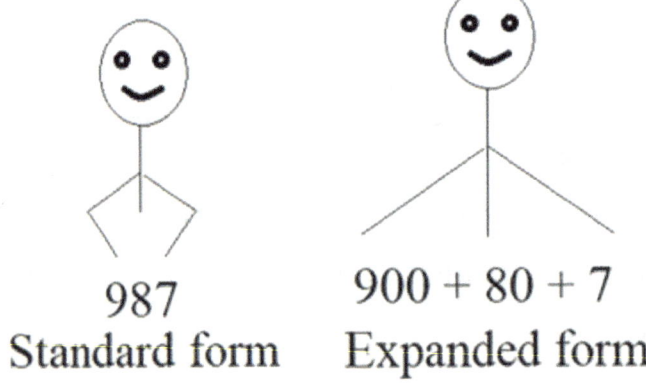

987
Standard form

$900 + 80 + 7$
Expanded form

e) Rounding Off

Underline the digit
Look right next door,
Five or higher,
Add one more
Four or lower,
Just ignore.

Example:
Round to the nearest ten
a) 376
b) 431

Solution:
Since we must round to the nearest ten, so we underline the tens digit and look at the ones digit.
a) 3_7_6 to the nearest ten gives 380 (ones digit is six which is higher than five, so we add one to the tens digit).
b) 4_3_2 to the nearest ten gives 430 (ones digit is two which is lower than four, so we just ignore it and put a zero).

Note:
i) To round to the nearest hundred, we underline the hundreds digit and look at the tens digit.
ii) To round to the nearest thousand, we underline the thousands digit and look at the hundreds digit.

f) Prime Number

PRime = I and ME (prime number has only two factors – one and itself).

Example:
17 is divisible only by 1 and 17.

g) Sum and Difference

i) Addition check

To check the addition, we must first understand the meaning of digital root.

Definition of digital root:

A digital root is a single digit number which is obtained by adding the digits of any number. If the sum obtained consists of two digits, it is added again. For example, the digital of root of 35 is 3 + 5 = 8 and the digital root of 48 is 4 + 8 =12, since it is a two-digit number the digits are added again. Hence the digital root of 48 is 3 (1 + 2 = 3).

1) To check the addition of two numbers, find the digital root of the two addends and the digital root of the sum (answer).
2) Add the digital roots of the two addends. If this sum is equal to the digital root of the answer, then the addition is correct.

Example:

2501 Addend (Digital root = 2 + 5 + 0 + 1 = 8)

+6289 Addend (Digital root = 6 + 2 + 8 + 9 = 25,

$$2 + 5 = 7)$$

8790 Sum (Digital root = 8 + 7 + 9 + 0 = 24, 2 + 4 = 6)

The sum of the digital roots of two addends is 8 + 7 = 15, 1 + 5 = 6 which is same as the digital root of the sum. Hence the addition is correct.

ii) Subtraction Check

1) Find the sum of the difference and the subtrahend.
2) If the sum obtained in step 1 is equal to the minuend, then the subtraction is correct.

Example:

8967 Minuend
− 4892 Subtrahend
4075 Difference

4075 Difference
+4892 Subtrahend
8967 Minuend

The sum of the difference and the subtrahend is equal to the minuend hence the subtraction is correct.

iii) Minuend containing zeros

To subtract a number from a three-digit multiple of 100, four-digit multiple of 1000 or five-digit multiple of 10000 first subtract 1 from both the minuend and the subtrahend. Then find the difference of the two results. This will give you the difference (answer) without regrouping.

Example:

        5000 Minuend containing zeros
       - 3968 Subtrahend

        4999 Subtract 1 from minuend
       - 3967 Subtract 1 from subtrahend
        1032 Difference

Hence 5000 − 3968 = 1032.

h) Product and Quotient

i) Multiplication Table of 3

1) Draw a 3 × 3 square grid, as shown in the following figure.

2) Write numbers from 1 to 9 starting at the
   lower left corner.

3	6	9
2	5	8
1 Start here	4	7

3) Write three zeros in front of the first row, three ones
   in front of the second row and three twos in front of
   the third row.

3	6	9	000
2	5	8	111
1 Start here	4	7	222

4) Insert zeros in front of the first-row numbers, one in
front of the second-row numbers and two in front of
the third-row numbers.

03	06	09
12	15	18
21 Start here	24	27

The table of three is ready. Just read it from top left to right in order. It's easy to remember 3 × 10 = 30.

ii) Finger Calculator

The following trick works for any multiplication of 6, 7, 8, 9 and 10 (it does not work for 2, 3, 4, and 5). For example, it will work for 6 × 8 but not for 6 × 4.

1) Number the fingers as shown in the figure below. You may write the numbers on your fingertip with a marker.

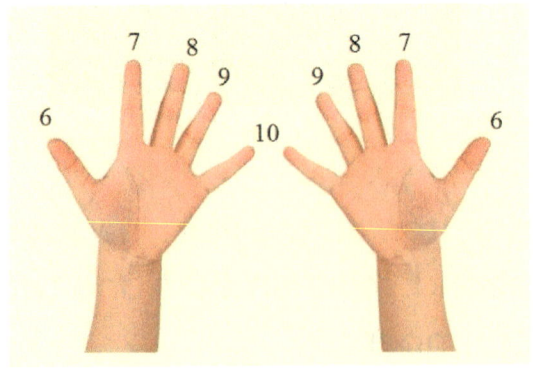

2) Suppose you want to find 7 × 6 then bend the fingers up to seven on the left hand and up to six on the right hand.

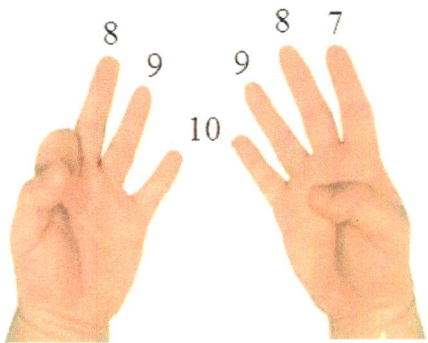

3) Each bend finger is a multiple of ten. In this case we have three bend fingers (two on the left hand and one on the right), hence we get 30.

4) Count the number of upright (straight) fingers on the left hand. In this case we have three upright fingers.

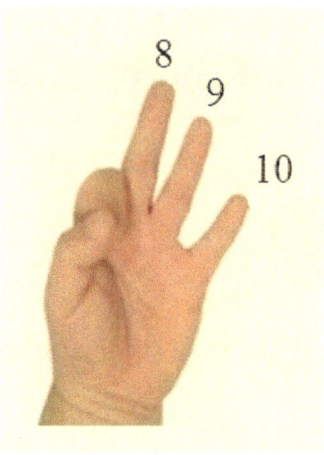

5) Count the number of upright fingers on the right hand. In this example, we have four upright fingers.

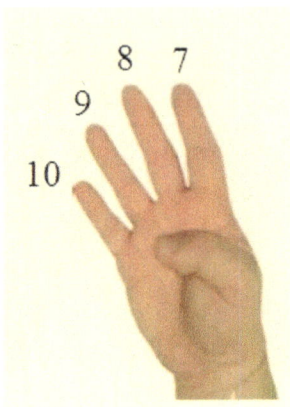

6) Multiply the upright fingers i.e., 3 × 4 = 12.
7) Add the numbers obtained in step 3 and step 6.
$$30 + 12 = 42. \text{ Thus, } 7 × 6 = 42.$$

iii) Zero Properties

a) Any number (say N) divided by zero = N ÷ 0 = Not defined (NO or not defined)
b) Zero divided by some number (say K) = 0 ÷ K = 0 (OK or defined)

i) Word Problem

To solve a word problem step by step, remember the word CUBES.
Circle the numbers and units (if any)
Underline the question asked
Box the key words (to determine the operation i.e., +, -,

×, ÷)

Evaluate (how to solve) and eliminate (words which are not needed)

Solve and check

j) Factor and Multiple

Factor = Fewer, Multiple = More (obtained by Multiplying)

Example:

Factors of 10 = 1,2,5,10 (Fewer)

Multiples of 10 = 10,20, 30,40, 50,60,70,80,90,100...(More)

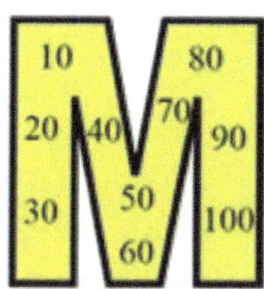 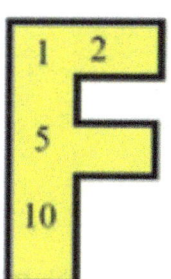

k) Fraction

Numerator and Denominator – To avoid the confusion of numerator and denominator, remember

Nice Dates $\dfrac{\text{Numerator}}{\text{Denominator}}$

Proper and Improper Fraction – It is proper for a big child to carry a small child, but it is not proper for a small child to carry a big child. Similarly, if the numerator is smaller than the denominator it is a proper fraction and if the numerator is bigger than the denominator it is an improper fraction.

Proper

Improper

I) Decimal

i) Place Value – In order to recall the decimal place value table, remember the sentence –

"Tim Hates Thyme & Tofu (Thick)".

Tim	Hates	Thyme &	Tofu (Thick)
Tenths	Hundredths	Thousandths	Ten Thousandths

To read a decimal number –
Read the number with no dot,

Then read the place value spot.

Example:

7.265 Seven and two hundred sixty-five thousandths

ii) Decimal multiplication and division by 10, 100, 1000

While multiplying a decimal number by 10, 100, 1000 the decimal point shifts to the right by 1, 2 and 3 places respectively.

While dividing a decimal number by 10, 100, 1000 the decimal point shifts to the left by 1, 2 and 3 places respectively.

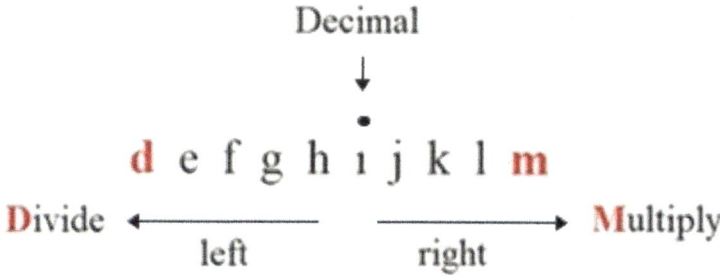

Example:

3.124 × 100 = 312.4

3.124 ÷ 100 = 0.03124

iii) Converting a repeating decimal to fraction

To convert a repeating decimal into fraction, write the number that is repeating in the numerator and that many nines in the denominator.

Example:
Convert the following decimals into fractions:

  a) 0.565656...56
  b) 0.847847...847

Solution:
a) $0.565656...56 = \dfrac{56}{99}$

b) $0.847847...847 = \dfrac{847}{999}$

m) Percent

i) Conversion of fraction to decimal to percent

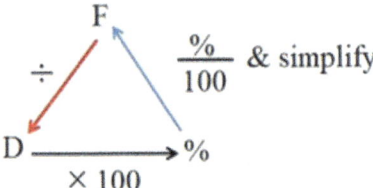

ii) Conversion of percent to decimal to fraction

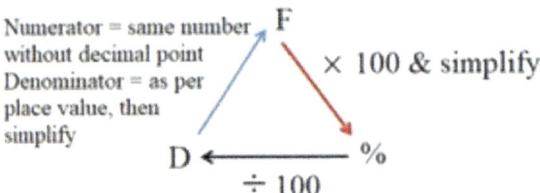

n) Integers

i) Addition of two integers
To recall the rules of addition of integers remember
SADS (same add, different subtract).
Same (sign) Add – If the signs are the same add the

numbers and keep the sign.

Different (sign) Subtract – If the signs are different, subtract the numbers and use the sign of the larger number.

Example:

a) + 4 + 5 = +9

b) – 2 – 3 = -5

c) –7 + 1 = -6

ii) Subtraction of two integers

To recall the rules of subtraction of integers remember KAO (keep add opposite) i.e., keep the sign of the first number same and add the opposite sign of the second number.

Example:

a) +9 – (–6)

$\quad$ = +9 + 6

$\quad$ = +15

b) –9 – (+6)

$\quad$ = -9 – 6

$\quad$ = – 15

iii) Multiplication and division of two integers

To recall the rules of multiplication and division of integers remember Solar Panel Dealer's Network i.e., same(sign) positive, different (sign) negative.

Example:
a) –2 × –3 = +6
b) +8 ÷ +2 = +4
c) +5 × –2 = –10

o) Geometry

Diameter and Radius

To remember the relation between diameter and radius remember dr.

In dr, 'd' is twice as tall as r. Hence, diameter = 2×radius

i) Angles

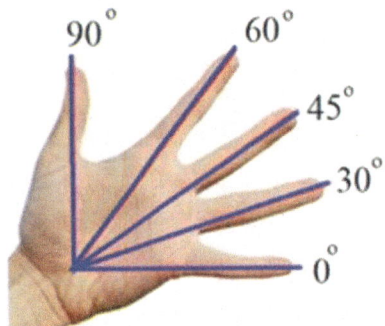

ii) Types of triangles

To recall the names of three types of triangles remember the sentence– "Eat Ice-cream Slowly"

Equilateral – three sides equal

Isosceles – two e, hence two sides equal

Scalene – not equal (No sides are equal)

iii) Angles made by parallel lines

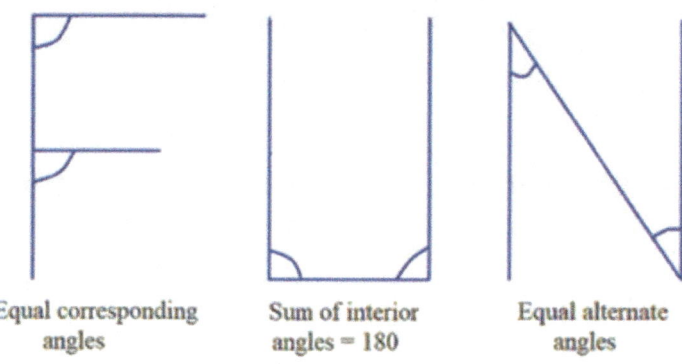

| Equal corresponding angles | Sum of interior angles = 180 | Equal alternate angles |

p) Calendar and Measurement

i) Knuckles

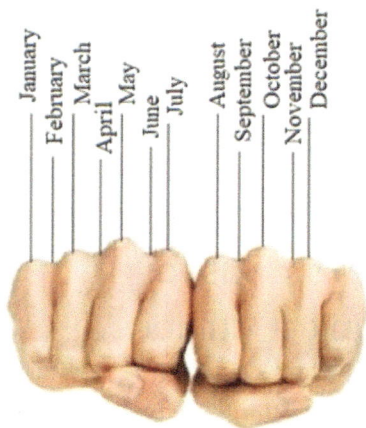

Make your fist and start on any knuckle with January. The space between the knuckles represents February. Continue counting by knuckle and the space between the knuckles. The knuckle is higher so the months on the knuckle have 31 days. The space between the knuckles is lower. Hence the months on these spaces have 30 days except February which has 28 or 29 days.

iii) Metric Measure

To recall the metric measure table, remember the sentence

"Kim Henry doesn't usually do craft & Maths"

Kim	Henry	doesn't	usually	do	craft &	Maths
Kilo	Hecto	Deca	(unit)	deci	centi	milli
Km	Hm	Dam	m	dm	cm	mm
Kg	Hg	Dag	g	dg	cg	mg
Kl	Hl	Dal	l	dl	cl	ml

vi) Changing units

a) To convert big unit to small unit remember
Best Soy Milk which stands for
Big Unit to Small Unit = Multiply

b) To convert small unit to big unit remember
Sweet Brownie Dessert which stands for
Small Unit to Big Unit = Divide

q) Area and Perimeter

Perimeter we can measure
Add all the sides,

Area is the treasure
Total space inside.

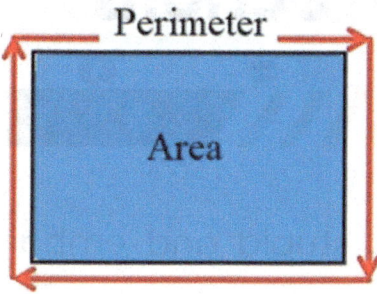

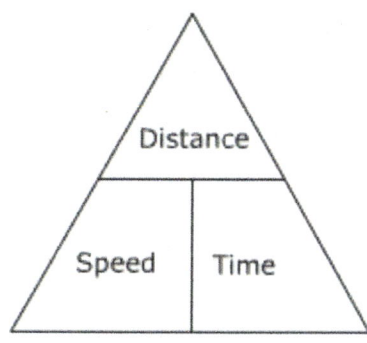

To recall the formula of distance cover distance with your finger so Distance = Speed × Time
To recall the formula of speed cover speed with your finger so, Speed = Distance ÷ Time
To recall the formula of time cover time with your finger so, Time = Distance ÷ Speed

s) Interest

To recall the formula of simple interest, remember "I am pretty".
I = prt, Interest (simple) = principal × rate × time

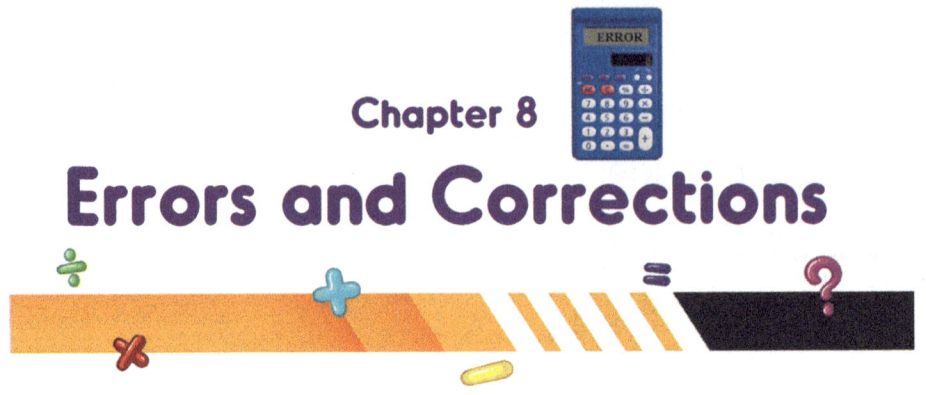

Chapter 8
Errors and Corrections

Mistakes can make us feel silly, stupid and embarrassed. It may develop a fear in our minds, but mistakes are not a bad thing. Every human makes mistakes. Mistakes are foremost part of learning. Mistakes/Errors does not reflect a lack of our ability. In fact, they are opportunities to learn new methods of thinking thereby increasing our knowledge.

Mathematical Errors:

There are mainly three types of errors in Mathematics.

1) Careless Error — These errors occur when the student does not pay full attention and tries to work quickly. It also occurs due to poor handwriting or trying to speed up unnecessarily.

2) Calculation error — These errors occur when any of the arithmetic operations (+, −, ×, ÷) goes wrong in any step. This occurs due to lack of practice or rushing through the problems.

3) Conceptual error – These errors occur when a concept is misunderstood by the student. It also occurs when a student applies the wrong logic.

Examples of common Mathematical errors:

Serial number	Error 😟	Correction 😊
1.	Fourty	Forty
2.	9th = Nineth	9th = Ninth
3.	Place value of 2 in 326 = Tens	Place value of 2 in 326 = 20 or 2 tens
4.	TO 64 +27 811	TO 1←Carry over 64 +27 91
5.	TO 90 −58 48	TO 81←Borrow 9̶0̶ −58 32
6.	326 + 10 = 3260	336
7.	9 × 0 = 9	9 × 0 = 0
8.	2 ÷ 8 = 4	$2 ÷ 8 = \frac{1}{4}$
9.	Three dollar five cents = $3.5	Three dollar five cents = $3.05

10.	 Fraction of Red rectangle = $\frac{3}{7}$	 Fraction of Red rectangle = $\frac{3}{10}$
11.	$\frac{1}{3} + \frac{1}{4} = \frac{2}{7}$	$\frac{1}{3} + \frac{1}{4} = \frac{4+3}{12}$ $= \frac{7}{12}$
12.	Three hundred and thirty-four thousandths = 0.334	Three hundred and thirty-four thousandths = 300.034
13.	0.2 + 1 = 0.21 0.2 + 1 = 10.2	0.2 + 1 = 1.2
14.	i) 8m 5cm = 85 cm ii) 5m^2 = 500 cm^2 iii) 3m^3 = 300 cm^3	i) 8m 5cm = 805 cm ii) 5m^2 = 50000 cm^2 iii) 3m^3 = 3000000 cm^3
15.	While using a ruler, measurements start from one.	While using a ruler, measurements start from zero.

Steps to avoid Mathematical errors

a) Concentrate on your work.

b) Read instructions carefully.

c) Write neatly.

c) Do not rush.

d) Check the solution.

e) Write all the steps.

f) Practice at least twenty problems every day.

g) Do activities based on Mathematics.

h) Solve Mathematical puzzles.

i) Play Mathematical games.

j) Do not hesitate to take help if a concept is not clear.

Chapter 9
Jokes

1) Teacher — What is the full form of Math/Maths
(expecting the student to say Mathematics)?
Student (in India) – Most Awful Tough & Horrible Subject
Student (in USA) – Mental Attack To Humans.

Note:
In Eastern countries like India the short form of Mathematics is Maths whereas in the Western countries like USA the short form is Math.

2) Teacher — How much is half of eight (expecting the student to say four)?
Student – Horizontally or vertically?
Teacher – Why do you ask that?
Student – If cut vertically it makes 3 and cut horizontally it leaves two zeros.

$$8 = 3$$
$$8 = \begin{matrix} 0 \\ 0 \end{matrix}$$

3) In the library:
Lily – Do you know where are the Mathematics books?
Marina – In the horror section.

4) Me doing Geometry:

Geo-me-try

Geo-me-cry

Geo-me-why

Geo-me-bye

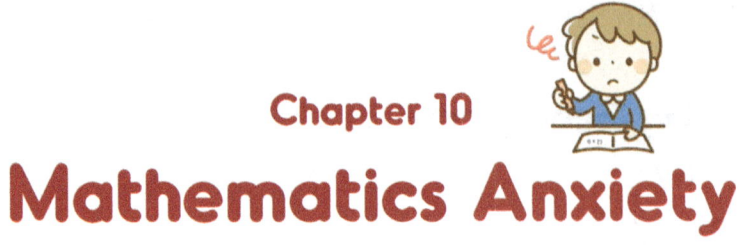

Chapter 10
Mathematics Anxiety

Mathematics anxiety is much more than a dislike for the subject — it's a real problem for the students. Psychologist Mark H. Ashcraft has defined Mathematics anxiety as "A feeling of tension, apprehension or fear that interferes with Mathematics performance".

Mathematics anxiety can affect students as early as grade 1. According to statistics 5 out of 10 students suffer from Mathematics anxiety.

Signs and symptoms of Mathematics anxiety

1) Negative thinking and negative self-talk

You often think of the following –
"I hate Maths/Math."
"Only smart kids can do Maths/Math. I can't."
"I can never ever do Maths/Math."

2) Avoidance

You may feel that studying other subjects is enjoyable but studying Maths/Math is highly troublesome. Thinking about Maths/Math brings such negative feelings that you just want to avoid it.

3) Lack of Focus

If your parents force you to do Maths/Math, you are just not able to concentrate. You are ready to do any other chore to avoid Maths/Math.

4) Lack of Response

During Maths/Math class if your teacher asks you any questions, your mind just goes blank. You are not able to reply and find it hard to think clearly.

5) Lack of self-confidence

Sometimes even if you know the answer, you prefer to remain quiet thinking that the answer might be wrong.

6) Physical symptoms

The day before the Maths/Math exam you tend to get a headache or a stomachache or feel like throwing up.

Your appetite decreases, and you are not able to sleep properly at night.

7) Unusual nervousness

During the Maths/Math exam you get sweaty hands, feel breathless, your heart pounds hard.

8) Tear/anger

All your feelings are real, though your parents or teacher do not seem to understand you. Probably they think that you are pretending. You feel helpless which results in anger or crying.

If you think you have any of the above symptoms don't delay act immediately.

Remedial measures

1) Ask for Help

If you don't understand any concept, ask your teacher for help the same day. The longer you wait the more difficult the subject will get. Asking for help is an essential part of learning. Your teacher will be glad that you have asked for help and explain the concept again to you.

> **Remember —** The struggle you feel today will offer the strength you need tomorrow.

2) Fight against your negative thinking

Do not think "Maths/Math is not for me, I can't do it, it is very hard". Use positive thoughts to replace your negative thoughts. Make positive thinking cards (you may use index cards).

Example:

Front of the index card	Back of the index card
a) Maths/Math is very hard.	a) Maths/Math may take some extra time and effort, but I will be able to do it.
b) Maths/Math is not for me.	b) Maths/Math isn't my best subject, but I can train my brain.
c) I always get bad grades in Maths/Math. I am stupid.	c) I will study harder next time and get better grades.

> **Remember —** It's okay to not know but it's not okay to not try. There is nothing impossible to do as the word itself says I'M POSSIBLE.

3) Do not panic

It is natural to make mistakes. Do not feel upset about it. Everybody makes mistakes. Mistakes are opportunities for the brain to grow.

Remember — Mistakes Allow Thinking to Happen (in Students)

4) Visualize success

Choose a quiet corner of the house. Be optimistic. Close your eyes and imagine seeing a movie in which you are the Maths/Math hero/heroine. Visualize yourself to be productive. Think of statements like – "I believe I can do Maths/Math, it's so easy."

See yourself winning Maths/Math contests and your teacher / parents appreciating you. You may see the movie for as long as you desire (just remember not to include any negative sentences or scenes). Now open your eyes and look for some Geometry shapes or different kinds of angles in your book. Close your book and draw the shapes or the angles in air with your finger (thinking it to be a pencil).

Remember — The famous quotations:

i) "Whatever you think that you will be. If you think yourself weak, weak you will be; if you think yourself strong, strong you will be".

– Swami Vivekananda

ii) "Believe you can, and you are halfway there".

–Theodore Roosevelt

5) Take a deep breath

Breathe in through the nose for a count of four, then hold the breath for a count of four. Breathe out through the mouth on a count of four. Repeat these at least three times.

Remember - Every day begins with a sunrise which is considered as a new beginning. Start your day with positive thoughts, do breathing exercise, eat a balanced diet, smile and start doing Math again.

6) Practice regularly

Even if you understand a new Maths/Math concept, unless you practice you will not remember. The basic key to success in Maths/Math is regular practice.

> **Remember —** "The only way to learn Mathematics is to do Mathematics".
>
> *-Paul Halmos*
>
> "Practice isn't the thing you do once you are good. It's the thing you do that makes you good."
>
> *– Malcolm Gladwell*

7) Use mnemonics

The greatest power lies in your mind.

> **Remember —** Use mnemonics to remember Maths/Math facts or techniques.

8) Solve puzzles

Puzzles develop spatial skills and make Mathematics enjoyable and fun.

> **Remember —** Even the hardest puzzles have a solution though it may take time to find it.

9) Play Mathematical Games

Mathematical magic /games help in grasping the Mathematical concepts and develop strategic thinking.

10) Make your own Math A-Z

This will help you in self-analysis and finding your weak chapters. Act immediately.

Say goodbye to Mathematics anxiety by following the above remedial measures and start loving the subject. After all Mathematics is an important part of our daily life.

Answers

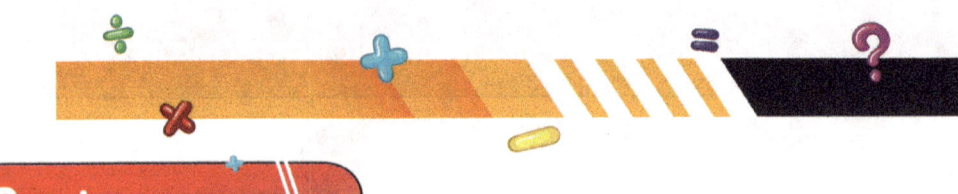

Puzzles

a) Geometriku

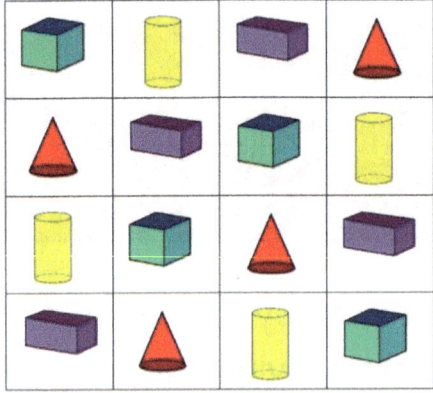

b) Shape number

If sum of three squares is equal to 12, then

one square = 12 ÷ 3 = 4

Similarly, if sum of three triangles is 9, then

one triangle = 9 ÷ 3 = 3

If sum of three circles is 6, then one circle = 6 ÷ 3 = 2

Thus, square + triangle + circle = 4 + 3 + 2 = 9.

c) Clock time

4:20.

The hour is increasing by one and the minute by five.

d)Three Numbers

The numbers 1, 2 and 3.

$1 \times 2 \times 3 = 6$, $1 + 2 + 3 = 6$.

e) Weight Measurement

They both weigh the same. Both are one pound.

f) Drying Clothes

One hour. The heat of the sun remains the same even for ten shirts.

g) Match Sticks

Remove the lower three match sticks and move them to the top

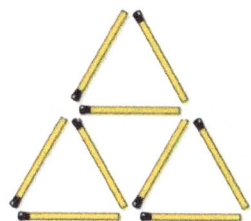

Five triangles (four small triangles and one big triangle).

h) Buying Stamps

She will get 12 (1 dozen = 12 units)

i) Shapes in flags

Shape	Flag and Country
Red circle	Japan Bangladesh
Red semi-circle	Malawi
Blue squares and rectangles	Greece
White stars (> 40)	USA
Green right-angled triangle	Namibia
Red trapezoid	Philippines, South Africa
Red parallelogram	Congo

j) Honeycomb

Honeycomb has a hexagonal shape because it needs

minimum bee wax for construction and can store maximum honey. For a given area, circle has the minimum perimeter followed by hexagon. But circular beehive would have resulted in unused space between the honeycombs. Hence, the beehive is hexagon in shape. It prevents any space from being wasted.

k) Cube

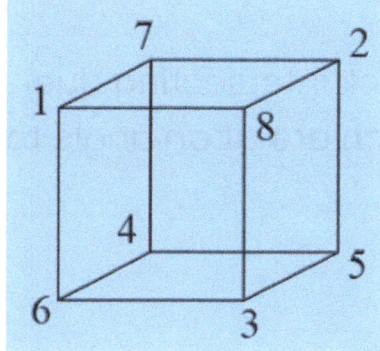

l) Digital Clock

10:08

m) Line Segment

Draw a longer line segment next to it. This will make the first line segment look shorter.

n) Line Sum

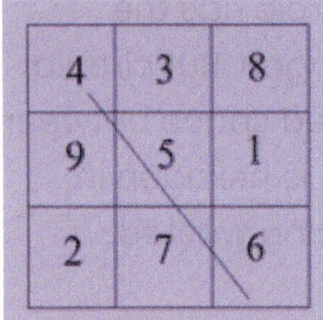

Optical illusions

a) Müller-Lyer Illusion

Both the lines are of the same length. The inward and outward arrows make them look different.

b) Zöllner Illusion

The thick lines are parallel. They look intersecting due to the background small lines which are at an angle to the thick big lines.

c) Kanizsa Illusion

A green equilateral triangle (pointing downwards) can be seen. There are no lines to indicate that triangle exists. Due to the circle and the black broken lines the figure looks like a triangle.

d) Hering Illusion

The lines are straight and parallel. The background spokes (radial lines) make it appear bowed at the centre.

e) Circle Illusion

Two big circles appear to be one circle when we take our nose close to the red (small) circle because our brain overlaps the two images.

Engrossing Patterns

1. a) $33334 \times 33334 = 1111155556$
   b) $77777 \times 99999 = 7777622223$

2. a) $99999 \times 2 = 199998$
   $999999 \times 2 = 1999998$

   b) $111111 \times 111111 = 12345654321$
   $1111111 \times 1111111 = 1234567654321$

3). a) $6/9 = 0.66666\ldots$
   $7/9 = 0.77777\ldots$
   $8/9 = 0.88888\ldots$

   b) $6/11 = 0.54545454\ldots$
   $7/11 = 0.63636363\ldots$
   $8/11 = 0.72727272\ldots$

4. a) $15873 \times 35 = 555555$
   $15873 \times 42 = 666666$
   $15873 \times 49 = 777777$
   $15873 \times 56 = 888888$

b) 15 × 37 = 555 and 5 + 5 + 5 = 15
   18 × 37 = 666 and 6 + 6 + 6 = 18
   21 × 37 = 777 and 7 + 7 + 7 = 21
   24 × 37 = 888 and 8 + 8 + 8 = 24

Activities

Measurement

The length of some body parts are close to each other.

Fraction, Decimal and Percentage

Total number of squares = 100

Serial Number	Color	Number of squares	Fraction	Decimal	Percent
1.	Yellow	6	$\dfrac{6}{100}$	0.06	6%
2.	Brown	26	$\dfrac{26}{100}$ $= \dfrac{13}{50}$	0.26	26%
3.	Blue	10	$\dfrac{10}{100}$ $= \dfrac{1}{10}$	0.10	10%
4.	Red	12	$\dfrac{12}{100}$ $= \dfrac{3}{25}$	0.12	12%
5.	Green	16	$\dfrac{16}{100}$ $= \dfrac{4}{25}$	0.16	16%

6.	White	30	$\dfrac{30}{100}$ $= \dfrac{3}{10}$	0.3	3%

Calendar

25 December 2175 will be a Monday.

Congratulations!

You have successfully completed the book -

"Fun with

Mathematics"

Other books by the same author

1. Title:Young Detective
Reading age: 9 years
and above

2. Title: Vedic
Mathematics
Reading age: 9 years and
above

3. Title: Learning Mathe-
matics the Fun Way
Reading age: 7-8 years

4. Title: 71 + 10 New
Mathematics Projects
Reading age: 10 — 14
years

5. Title: Excel with Mathematics Pre Primer
Reading age: 3-4 years

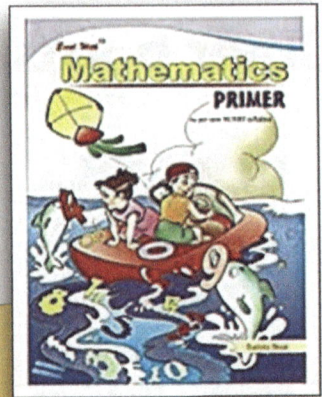

6. Title: Excel with Mathematics Primer
Reading age: 4-5 years

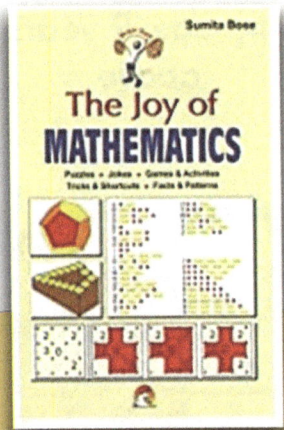

7. Title: The Joy of Mathematics
Reading age: 10 years and above

8. Title: Autism – A Handbook of Diagnosis and Treatment of ASD (This book created a record and was entered in "The India Book of Records 2016". They referred to this book as "First book on autism in Indian socio cultural set up")
Reading age: For parents, caregivers and teachers

All the above books are available at Amazon, Flipkart and various book stores.

Made in the USA
Monee, IL
07 July 2026

56548179R00050